WEIRD SPACE SCIENCE

RADIATION IN SPACE

VIRGINIA LOH-HAGAN

45TH PARALLEL PRESS

Published in the United States of America by Cherry Lake Publishing Group
Ann Arbor, Michigan
www.cherrylakepublishing.com

Photo Credits: © NASA images/Shutterstock, cover, 1, 16; © Shivam Chy/Shutterstock, 6; © NASA, 8, 21; © NASA Goddard/NASA, 10, 21; © Illustrator76 / Shutterstock, 13; © Juergen_Len/Shutterstock, 14-15; © NASA images/Shutterstock, 18-19

Graphic Element Credits: Cover, multiple interior pages: © WindAwake/Shutterstock, © kichikimi/ Shutterstock, © Your Local Llamacorn/Shutterstock, © Marishiav/Shutterstock, © Supza/Shutterstock, © HSSstudio/Shutterstock, © Olive Kitt/Shutterstock, © KKarance/Shutterstock, © Theo_hrm/Shutterstock, © Rodin Anton/Shutterstock, © Petr Vaclavek/Shutterstock, © Nikolaeva/Shutterstock, © Evgenii Doljenkov/Shutterstock

45th Parallel Press is an imprint of Cherry Lake Publishing Group.

Library of Congress Cataloging-in-Publication Data has been filed and is available at catalog.loc.gov

Cherry Lake Publishing Group would like to acknowledge the work of the Partnership for 21st Century Learning, a Network of Battelle for Kids. Please visite Battelle for Kids online for more information.

Printed in the United States of America

Dr. Virginia Loh-Hagan is an author, university professor, and former classroom teacher. She's currently the Director of the Asian Pacific Islander Desi American Resource Center at San Diego State University. She loves science-fiction stories. She lives in San Diego with her very tall husband and very naughty dogs.

TABLE OF CONTENTS

BREAK IT DOWN

1. DECODE IT

- Look at each letter.
- Look at letter patterns.
- Blend sounds.
- Try other possible sounds.

2. BUILD VOCABULARY

- Listen to the word.
- Match the sounds to a word you know.
- Look for context clues for words you don't know.
- Look up the meaning in a glossary or dictionary.

3. MAKE CONNECTIONS

- Name other examples.
- Draw the idea.
- Compare ideas.
- Ask questions.

TEST IT OUT

WORDS YOU'LL LEARN

Look over the list of vocabulary words. Are there any you already know?

galaxies
light waves
limited
radiation
reflects
theories
visible

GET STARTED

Open and closed syllables are 2 types of syllables. Open syllables end with a long vowel. Closed syllables end with a consonant. They have a short vowel sound.

- Look at the letters.
- Divide words into syllables.
- Try dividing another way if needed.

The word *galaxies* starts with a closed syllable. It is pronounced: GAL-uhk-seez.

- The 1st vowel is a short *a*: *gal-*.
- The 2nd vowel says "uh": *-uhk-*.
- The 3rd vowel is a long *e*: -seez.

Do you know why space is black?

Scientists studied this question.

They studied pictures and sounds.

They came up with **theories**.

These are explanations. They explain why things happen.

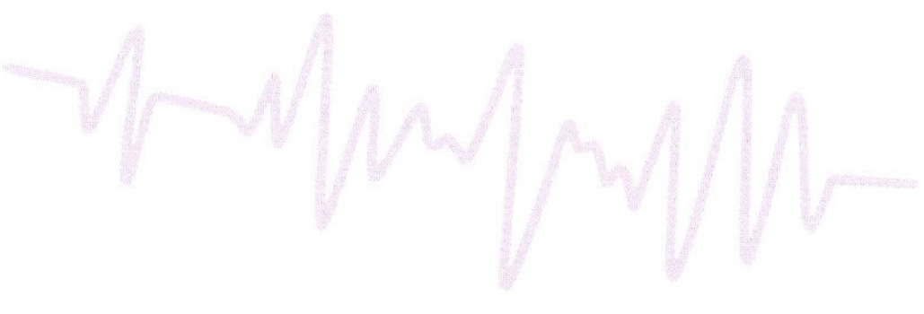

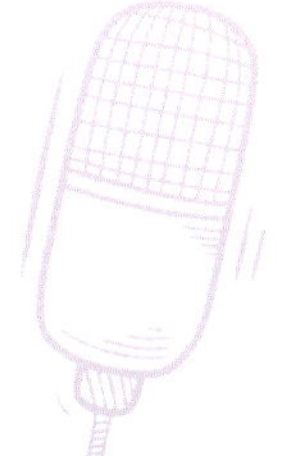

Heinrich Wilhelm Olbers was a German scientist.

He said if space went on forever, stars should fill the sky. Night would be bright.

He had a theory. He guessed there were a **limited** number of stars.

Limited means having an end.

DECODE IT

The word *limited* starts with a closed syllable.

- Break it into chunks: lim-i-ted.
- The 1st vowel sound is a short *i*.
- The other vowel sounds say "uh."
- *Limited* is pronounced LIM-uh-tuhd.

Visible light is light we can see.

The universe is filled with light. Much is light we can't see.

Scientists study **light waves**.

These are how light travels.

Light moves in a straight line.

BUILD VOCABULARY

Reflects (rih-FLEKTS) means bends back.

- Light does this. Sound does this.
- A mirror reflects what you see!
- Other forms: reflect, reflection, reflective, reflector

It **reflects** off something. Reflects means bends back.

Or it's bent by a lens.

The Moon reflects the Sun's light. It bends the light towards Earth.

Most of outer space is empty.

Light is **absorbed**. Absorbed means taken in.

This means we only see black.

Light is a type of **radiation**. Radiation is energy waves.

These waves are in a pattern. They go from long waves to short waves.

We can only see a fraction of waves. We see them as visible light.

Light from other **galaxies** often has long waves. Galaxies are big groups of stars.

DECODE IT

The word *radiation* has 3 open syllables.

- Break it into chunks: ra-di-a-tion
- The 1st and 3rd vowels are a long *a*.
- The 2nd vowel is a long *e*.
- The last vowel sound says "uh."
- *Radiation* is pronounced ray-dee-AY-shuhn.

Light waves stretch out as the universe expands.

They become too long to see.

That is why space seems black to us.

MAKE CONNECTIONS

When we see light in space, it is from the past. Most of space seems black.

- What objects in space give off light?
- Think about light in space. Why would the light we see be from the past?

Scientists can use special tools.

They can see radiation that our eyes can't!

PUT IT TOGETHER

Detail:
Scientists of the past studied why space seems black. They had different theories.

Detail:
We can't see most radiation waves.

Detail:
Some scientists investigate light waves.

Main Idea:
Space looks black to us.

Detail:
When light waves expand, they are no longer visible to humans.

Detail:
Light waves expand as the universe expands.

LEARN MORE

BOOKS

Loh-Hagan, Virginia. *Introduction to Space*. Ann Arbor, MI: 45th Parallel Press, 2025.

Loh-Hagan, Virginia. *Weird Science: Space*. Ann Arbor, MI: 45th Parallel Press, 2022.

Tyson, Neil deGrasse. *StarTalk: Young Readers Edition*. Washington, DC: National Geographic Kids, 2018

ONLINE

Search these online sources with an adult:

- NASA | Games and Interactives
- NASA+ | Explore Topics

GLOSSARY

galaxies (GAL-uhk-seez) huge space collections made of billions of stars, gas, and dust

light waves (LIYT WAYVZ) waves of light; how light travels

limited (LIM-uh-tuhd) having a limit or end

radiation (ray-dee-AY-shuhn) waves of energy sent out by light or heat; waves sent out from radioactive material

reflects (rih-FLEKTS) bends back from a surface

theories (THEE-uh-reez) understandings based off of scientific facts; they explain why things happen

visible (VIZ-uh-buhl) able to be seen

INDEX